THE DIGITAL CHURCH FIELD GUIDE

The Art of Digital Discipleship:
How to Expand Your Church and
Reach People Online

2023 - 2024 Edition

Cover Design by Luke Gajary

Author Photography by Double Knot Photography

Special discounts are available on quantity purchases. For details, contact me at luke@digitalchurch.co

This book is a time-sensitive resource and is meant to begin a larger conversation of reaching people online. Visit us at DigitalChurch.co to continue the conversation.

ISBN 978-0-9986620-3-9

CONTENTS

INTRODUCTION

Jesus had a media team.

Matthew, Mark, Luke, and John **recorded** His works in the media of the day – the written word.

He had a team of volunteers who were in the trenches with him, documenting what was happening.

If Jesus did not have a team around him whose main job was to capture and re-tell the stories of what was happening, a huge chunk of the New Testament would be missing.

While many of us are sometimes resistant to new technology, or using the tools available to us to reach others, the truth is that Jesus had people around him capturing messages, writing them down, and sharing them with others.

My name's Luke Gajary, and I'm the founder of an organization called DigitalChurch. I'm a digital marketing expert and a licensed minister, passionate about reaching people for Jesus and helping the body of Christ win in the online space.

I talk about that phrase a lot: "The church winning online", and what I mean by that is the church that wins online is simply the church that's **reaching people online**.

Too many of us play the game of keeping up with the latest megachurch down the street.

We steal stage ideas, sermon ideas, structure elements, ministry handbooks - and all that is fine. Nothing wrong with emulating people that are doing things we like (well, to a degree - I'll talk more about that later).

The danger in that though is losing your church's authentic voice in an effort to be like "Random Action Verb" Church.

Austin Kleon, in his book Steal Like An Artist (one of my personal favorite manifestos on creativity) wrote "You don't need to be a genius, you just need to be yourself."

"You don't need to be a genius, you just need to be yourself"

That rings so true and needs to be said in our relentless pursuit of "keeping up with the Jones' " that we give in to.

I don't believe that our numbers need to compare to everybody else's numbers. I don't think you have a certain "online church" benchmark you need to be hitting as you refine your focus as a pastor, and as a leader.

> Our main goal should simply be reaching the specific region or group of people God has called us to the best that we can.

As pastors, we're called to reach people, but we didn't always understand the depth of what we were getting into when we accepted that call. One unforeseen side effect of the COVID-19 pandemic was that it forced us to all become instant digital media experts. A lot of us went from zero to 100 when it comes to the online space in our churches… all of a sudden we had to ramp up to a hundred miles an hour out of nowhere with zero to little training.

I haven't talked to many Bible Schools or seminaries that offer digital media programs to their pastors.

I have never met a pastoral search committee that pre-screens for extensive digital marketing background as a prerequisite for lead pastoring.

You're a pastor called to serve and love people, but you're now also in the difficult position of carrying the weight of digital media and online engagement for your church.

In this simple book, I want to zoom in on a few areas to help. In a sense, this will serve as simple online audit of your church - a field guide of a few best practices..

There are a couple of quick things that you can do to level up your church's online game that don't require a lot of budget or any board meeting decisions. The goal is by the end of this book that you will implement two or three things that can make your church better online.

This book starts off by assuming that you already have somewhat of an online presence, either you've got some social media, you have a website, or you have something going on. We're not starting from the ground up here – you may not be 100% happy with what you do have, but this tool is for those of that already have some online things established. I want to walk you through a couple of things you can do to make those things better, and then I'm going to introduce a new strategy that maybe you haven't thought of, or maybe one that you haven't thought of enough.

Each section will conclude with a three minute action item checklist that you can implement. This book isn't meant to just be informative, but to to be a field guide for improving your church's online presence.

It won't change everything overnight, and tech fairies won't emerge from the pages to fix all your online problems (we tried capturing some tech fairies, turns out they're very elusive) - but you will finish this book with some tangible, actionable steps.

3 minute action item:

Define your win. The difference between little league and MLB players? One is playing to have fun and the other is playing to dominate. What is your "domination?" What is the territory God has given your ministry to reach, and how can an online ministry help do that? Are views important? Using social media to get more people in the door?

THE BILLBOARD YOU'RE NOT USING

The first thing I want to highlight is your church website. Almost all of us church leaders have a website of some kind, and I'm going to challenge you here to really dig into your website and make it the best version that it can be. Not a full overhaul or redesign - unless there's any Comic Sans or Papyrus on it…in which case put this book down and fix it immediately.

Otherwise, I'm just going to suggest you do a couple of things today to change your current church website. It's absolutely amazing to me how good most church websites have gotten. Most churches have amazing websites (insert my terrible Trump impression here). They're beautiful - terrific. Huge!

Website-building technology has come so far since I was first introduced to the internet in the mid-'90s, and it has really been amazing what churches have been able to do.

However, I also see a lot of churches miss this one key concept on their site.

The most important thing for your church website, at least on the homepage, is to feature…time and place. I could almost hear you from beyond the pages of this book saying "Well, duh!".

I know, you would think it's pretty straightforward! I recently went on an "online church tour" of websites in 6 different states. I perused 28 different church websites of all different sizes. Out of those, 23 of them DID NOT CLEARLY SAY when and where they meet. Only 5 clearly stated times and places clearly for a quick visitor!

When people are first landing on the homepage of your website, do they see where and when you're meeting? Do they see clear information? A phrase I'm going to introduce to you is “above the fold”. What I mean by that is on a desktop computer, when you load the website, in that initial visible square of the screen that when you first land on the website, is your information right there?

Do they have to click three levels deep to find where you meet and when you meet? Do they have to search for your basic information? Of course there are a few different schools of thought when it comes to online web development and church websites, and I’ve had conversations with some “experts” who disagree with this.

I believe your most vital information, people should be able to find you quickly just by landing on your homepage. And this includes a mobile device, so that narrower screen should be the first screen that they see. Usually, there’s a big “hero” or header image first- that’s a big image of your pastor or a staff member or your church body. Overlaid on top of that is simple text that says what city you’re in, your address and your service times, and you could even include streaming platform info. The important thing here is that on mobile or desktop, it needs to be above the fold.

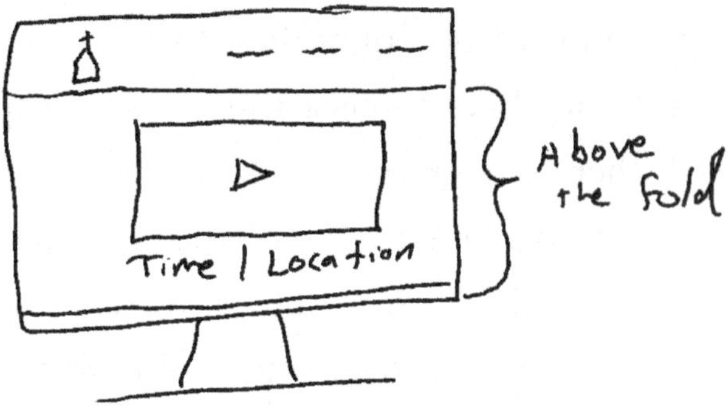

Bottom line: people should not have to scroll very far for that basic information. And time and time again very large churches, even what I would consider very advanced churches in the online space miss this. It's like we prioritize the coolness of our website over the simplicity of the information.

YES, make your website look cool. I've built and designed dozens of church websites, and I strive to make them all the coolest, most innovative church websites that ever existed. But if the basic information isn't there and if people have to search to find out when we meet as a church, there's a problem. Take a quick audit of your website – go to your church homepage.

3 Minute Action Items:

1. Spend 30 seconds on your home page - that's it. Evaluate if all the information a guest needs is "above the fold". Do they have to scroll to find location and service times?

2. If the answer is yes, email your web guy or log in and do it if you are the web guy or girl. Get vital info quickly accessible.

Easiest, quickest quiz you'll ever take - it's pass/fail: Do you have to scroll to see time and location? Do you have to click anything to see it?

If the answer to either of those is yes, put this e-book down and go change it. Email your web developer or staff member something like this: (copy and paste if you'd like and just edit):

"Hey,

Want to make sure our service info is pretty easy to spot on the website. Can you add Sundays at 10am, and our church address, to the header image of the homepage? Thanks! Also you're getting a raise!

- Pastor Jim"

It should take three to five minutes for you or your web guy to jump into your website and make those few changes.

LINGO AND LINKS: THE SOCIAL MEDIA BIO

The second thing I want to talk about is your church's visibility and social media presence. Similar to the website's "above the fold" concept of the basic information being readily accessible, I want you to remove as many layers as you can for people to find information on your church.

It should be as easy as possible for them to connect with you on social media. I'm not saying your Instagram feed has to be the most amazing thing by the end of this book. Don't stop everything you're doing and instantly go post 50 things on social media so it looks awesome.

There are a few simple things we can do today to make it look better overall, and to make it easier to access for the average person who stumbles onto your church's social media.

First, let's talk about the bio. Time and time again, I see churches advertise themselves. They're talking about the origin story. They're talking about "how we started, what our vision is".

That's all fine to talk about. A lot of churches have great God stories and incredible vision.

But at the end of the day, what's essential to the person visiting?

Think about the age-old marketing concept of what the customer is constantly asking: what's in it for me? What am I getting out of this? Don't just tell the story of your church. Tell the story of what the person coming onto your social media is looking for. How do you help people? What do you do for people? What do you do that matters to the person?

I am NOT SAYING to bury or ignore the story of your church. Your story is incredible and it should be out there.

But think for a moment about the single Dad who has been out of church for 8 years, and wants a safe place to bring his kids and get connected back into a Godly community of men. For this guy, say, Jeff, stumbling into your social media, his coming into something that's just advertising your church? Is it just look at us, all about us, us, us, us? Or is he going to see things that speak to him?

A common phrase in the IT or app development space is "the end user". We are making this for the end user.

Does your social media bio consider the end user, "Jeff"? Is it about that person or is it about you?

Make it less about you and your vision, and make it more about what is in it for that person.

Only you can come up with that. I can't write that for you. I can, if you want to talk after this, we can connect and brainstorm some ideas for visitor engagement. You've got to make it fit and feel natural.

Here's a few that I really like:

Timber Creek Church:
Your God-given Potential is Our Mission

Revive Church: **Introducing Real People to the Real Jesus**

North Point Community Church
Life is complicated. You want to get it right. We're here to help.

Vivid Life Church: **Pertenece y se parte de una comunidad auténtica de fe que han decidido crecer juntos y por consecuencia impactar a nuestra comunidad.**

(Belong and be part of an authentic community of faith that has decided to grow together and impact our community.)

The bio should make me as a visitor to your social media feel like this church is going to change my life. I need to know the story of your church, but the bio of your social media is not the place for that. It's about me. It's about the customer, the visitor. I'm not saying your church visitors are customers, but I'm using online marketing lingo to help you to think more like a digital marketer. **The best marketing is driven towards the customer, not the business.**

Another thing, while we're in the bio, let's talk about accessibility to your church online. A great tool I have used for years is called Linktree.

There are other services that do the same thing, and most of them are free.

What they do is create a series of buttons that link to your social media. For the average user on a phone, they're clicking on the link in your bio and it's trying to load your entire church's website. Sometimes that's a lot. What Linktree does is gives you five or six buttons that you can add content to, and it very quickly pops up with very little load time. It's a very minimal way for a user to quickly access certain things.

You'll want to think about what's priority to your church for your first time visitor, and link those things in that bio. Is it watching sermons? Is it giving? Is it directions to your church? Is it the latest initiative that you're doing? The latest campaign? What is it that people need to know about? Find those five or six things, reduce them down to buttons on Linktree, and then you can change those out as you need. So if you've got an event coming up, an outreach, or a new sermon series, you can switch those out seasonally and make your bio very relevant and accessible. Again, there's other services out there like this, but Linktree is an easy to use and probably the most popular one.

Now moving past the bio, another thing you should be focused on is good photography. This one's going to take a little more effort but it doesn't have to be intimidating. The ideal scenario for most churches (besides hiring a local freelance photographer) is a college student with a DSLR who can take really good photos of your services on a semi-consistent basis. A good shoot once a month or once a quarter can really help update photos and give you an archive of things to pull on.

I'm a firm advocate of the idea that social media feeds should look more like photos of your people and not just designs of your sermons or events. Your designs are awesome, I'm sure…but what's going to pull people in is pictures of faces, pictures of people! Interactive, smiling, friendly people.

Good photos are a big deal. Even if this just means the latest iPhone that somebody in your family has, if they can just take a few good photos. Lightroom is a great app to edit them. There's other awesome apps that you can use to simply edit photos and put them online.

Any photo is better than a stock photo. I say that hesitantly, because I know some people are capable of taking really bad photos, but I just think authenticity wins when it comes to these photos. It needs to be authentically your church. It needs to be your people. Even if they're not awesome to start. You'll get awesome eventually, but you've got to start using photos and not just designs and graphics.

Graphics are great. There's a lot of cool things on YouVersion and a lot of cool resources that make great graphics, but ultimately, when I land on your social media, I want to see photos of your people and photos of your church community.

Another key to consider is your header graphic when it comes to Facebook, YouTube, Twitter, or any platform that allows you to have a header graphic on your profile.

Similarly to when I talked about your website, your social media header provides an opportunity for somebody to find where you are and when you meet.

Different social media platforms offer different sizes, so google "Facebook header size", and use a free app like Canva or CC Express to create this header. Add your logo, add some cool photos and add some text. It's super simple to do, and you can update it and create it all straight from your phone. You're going to have to do a little bit of work in figuring out the right dimensions for each platform.

Additionally, make sure that the information is smaller in the center, because it's going to get cropped down on mobile versions. I've made the mistake before of designing graphics that looked great on the desktop, but when you pull it up on a mobile phone, everything cropped way out of frame and it looked awful. Test it. Don't be afraid to delete and re-upload your social media presence.

3 Minute Action Items:

1. Review your bio or about section on Facebook and Instagram. Update location and service times if they're missing. Communicate to the heart of the person scrolling.

2. Set up a free LinkTree for the bio link and add 3 top links.

3. Brainstorm 2-3 people in your church right now that you can test with taking photos of services once a quarter and make a plan to reach out to them

VIRAL AND VIRTUAL OUTREACH - REELS AND SHORTS

"It's weird. I haven't heard a single person talk about it. This is the first time where content can actually go viral on every single platform, and you don't have to film a video for Facebook film a 12 minute video for YouTube film, a sub 62nd video for Tik Tok, write a tweet for Twitter and post this on Reddit. You can just do the same thing on every platform."

Jimmy Donaldson, AKA Mr. Beast, top YouTuber in the World, in winter of 2023.

1. According to a report by eMarketer, in 2020, the average US adult spent 26.9 minutes per day watching short-form video content on their smartphones,

which was up from 23.9 minutes in 2019. (Source: eMarketer)

2. TikTok has become one of the most popular apps for short-form video content, with over 1 billion active users worldwide as of February 2021. (Source: TikTok)
3. In a survey by Wibbitz, 58% of Gen Z and millennial respondents said they preferred watching short-form vertical videos on mobile devices over other formats. (Source: Wibbitz)
4. Instagram Reels, a short-form vertical video feature launched in August 2020, has seen significant growth in usage. According to Instagram, as of January 2021, over 90 million accounts had created Reels. (Source: Instagram)
5. In a survey by HubSpot, 85% of businesses said they use video as a marketing tool. Among those, 38% said they specifically use short-form vertical

videos for marketing purposes. (Source: HubSpot)

6. In a study by Snapchat, 71% of Gen Z respondents said they prefer using their smartphones for content consumption over other devices, and 81% said they enjoy using vertical video formats. (Source: Snapchat)

These statistics demonstrate the growing popularity of short-form vertical video content, particularly among younger generations. The rise of apps like TikTok and Instagram Reels, as well as the widespread use of vertical video for marketing purposes, suggest that this trend is likely to continue, if not accelerate, in the coming years.

Businesses small and large are playing catch up to the video trends - building offices with in-house studios, hiring specialized video content creators, and creating content creation strategies within their marketing.

Short-form vertical video has not been in the cards for many organizations, and they are taking a ground-up approach to building it in now as an add-on to their strategy.

Here's why this is a big deal for ministers like us and the advantage that churches have - you already have an hour of content every week to pull from! This is not a new asset that needs to be created, a new team to be built, or a new department to be formed. We've just got to leverage the existing livestream service asset, reformat, and distribute it across social media to reach people.

Something great that came out of the Covid pandemic was it forced us all to catch up to speed with online streaming, get our churches good camera gear, and get syndicated across various live-streaming platforms.

The problem is that it leaves us with a participation trophy of sorts. We feel like we've reached our creative goal simply by streaming our content and making the full service available to access online all week.

Streaming services online used to be the height of digital media for churches. Now, it's the bare minimum, and using Shorts, Reels, and TikToks is the most effective organic media strategy.

We have all heard the adage that in sports "The best defense is a good offense". The thought is that by being proactive rather than passive, you will get a strategic advantage over your opponent by forcing your opponent to be on the defense.

The quote is attributed to Sun Tzu, the Chinese military general, writer and philosopher who in his book The Art of War wrote that "Attack is the secret of defense; defense is the planning of an attack". No matter the source, the notion of going on the offense to push back against threats applies to many cases and has broad application to war, games, business, and even life in general.

Does this apply to church media?

Many of us operate defensively in our online media approach. We've put the content up, we streamed our service, and now we are waiting to be approached.

It's time to go "on the offense." Package content into an easy-to-digest format and deliver it to the places our unreached audience is.

Here are four good reasons to start leveraging your sermon videos into shorts and reels as soon as possible:

1. You already have the source material.
The sermons you produce weekly and spend hours on research and development can be used more effectively. You don't have to develop a new separate teaching or series for social media.

2. It's an affordable and effective way to reach people outside your regular weekly service
Based on actual data I've experienced, most Reels and Shorts get exposure to non-followers. Think about the one sheep vs. the 99 – short videos are a great way to reach out to the "1" while simultaneously serving your "99" in the congregation.

3. It stretches your Sunday message through the whole week and beyond.

Have you ever felt like a message you preached was just dynamite? You put your all into it, and then Sunday at 1 pm…it's just over. Shorts and reels help capture the sticky thoughts of your message and amplify the key ideas further. Due to the algorithms' nature, many videos will appear in users' feeds for several weeks after they are posted.

4. Social algorithms are pushing this content right now - strike while the iron is hot.

The nature of social media changes very quickly, so while video content is the priority of social media platforms right now, why wouldn't we leverage our messages into that delivery system?

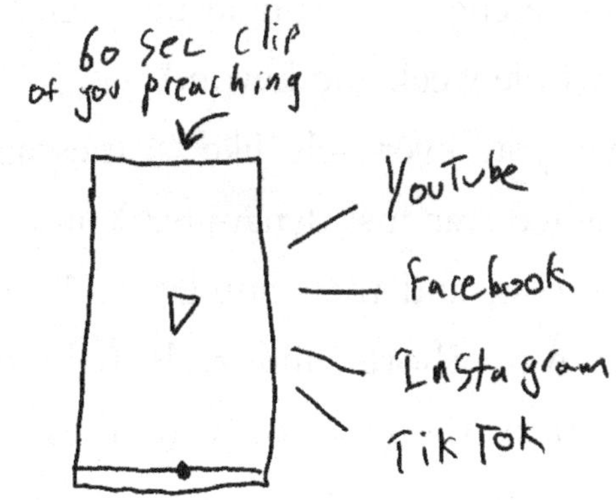

So what should these videos be? Can we throw any random clip online?

Marketing principles teach us a practical recipe for attracting people to our message are these three main ingredients:

1. Hook - an angle that fuels curiosity
2. Lesson - Teaches a new skill or value point
3. Emotion - Relatable storytelling

Use these three keys when creating short-form video content; your users will flock to and share your videos.

Jesus obviously didn't use video, but He used those same principles to teach the crowds he ministered to—a controversial statement: thought-provoking hook. Teaching about how the Kingdom worked: Lesson and value add. Parables: relatable and accessible stories.

Social networks will come and go, algorithms will change, and content delivery systems will be overhauled, but the principles of storytelling and communicating truth to transform hearts will never go away.

3 Minute Action Items:

1. Create a plan to start re-purposing your sermon content. Start with one clip per week.

2. Download CapCut or Premiere Rush to help edit videos on the phone

3. Find a member of your media team who would be willing to add this to their workflow and get them some resources to get started. I have some resources at digitalchurch.co that walk through some of these principals and give some best practices on getting started.

CAN'T TALK RIGHT NOW, CAN YOU TEXT INSTEAD?

I went to the Promised Land last weekend. A land flowing with milk and honey. Okay, so in reality, it was a Cheesecake Factory in north Dallas. There was a 15 minute wait for a table, so I asked to be put on the list. They required a phone number to text me when the table was ready. I gave it freely - because I wanted what they had. I didn't say "Well, I'd like to see a list of your core beliefs as a restaurant before I give you my number". "I'd love to talk to your manager about his view of eschatology before agreeing to be texted." "Wait, are you going to use this number just to ask me for more money?"

People that are connected to your church WANT to hear from you. If you are offering something of value, they will sign up. If you have nothing of value to offer them, they won't.

My last focus to help level up your church online is to set up a texting service. And this one, I will confess, is going to require a little bit of budget and setup time. There are some very, very affordable options, and there are some companies that will work with smaller churches and church plants to help scale up, for those of us on a budget that maybe we need to start somewhere and grow into it. There are options available for that.

"Well, we send out an email blast. Why do we really need a text message? Do people really want to hear from us as a text message?"

Speaking both from marketing analytics here and personal experience as a church creative: texting is so much more valuable than email.

Keep the email; I'm not saying ditch email. I think you have to reach people in their inboxes too. But texting... I mean, the statistics are incredible.

90% of text messages are opened within three minutes.

Three. Minutes.

I've done email marketing for a long time. Just for comparison, a really good email open rate is 15-25%. And that's just over the course of a week, that's not even saying within minutes. That's saying within a few days on a good day 15-20% of our readers are opening this email.

I was on staff at a church years ago and our open rate was 38% on emails. It blew me away - the best open rate I had seen in a while.

With text messages, 90% are opened within three minutes. We should be taking advantage of texting.

I'm willing to bet your people will not be mad about it…If you do it tactfully (so don't blast them more than once a week). In the way that I said that Reels are a good way to reach non-followers (the 1 sheep), texting helps you tend the 99.

That said, don't pummel them about every event. Choose what's important and send out details of the information that's relevant to serve and connect with them. Add value to them. Share things that are helpful, share things that are valuable and don't overload them.

I don't think people will be mad that their pastor or church staff is texting them about church events. Through my years of being on staff at multiple churches, the number one feedback we received was that there was a communication breakdown - they didn't feel communicated to enough. No matter the church or position I was in, communication with congregants did not always happen effectively. People want to feel connected and in touch with what is happening. If you don't abuse it, it will be embraced.

Clearstream is a service that I've used and set up and have helped a lot of people connect with. It's very easy to use and it's fairly affordable. There are different plans on their website, Clearstream.io. It's got some different scaling options for church planters and smaller churches, and free trial plan so you can try it out.

One of my favorite features about Clearstream is that it actually synchronizes with Planning Center (church database software). If you don't use PCO but use another database, it has easy ways to import and export lists. You could export the phone numbers out of your database and import them into Clearstream. And for those of us with Planning Center, it automatically syncs lists and member data within Clearstream.

I can instantly pull a list out of Planning Center into Clearstream and text that list.

The way most of these services work is that you buy credits. Every text message with 160 characters or less costs one credit, and then you use up a credit every time you send a message. So if I send a message to a list of 50 people, that costs me 50 credits. You buy a certain amount of credits a month, and once you burn through them, you have to either buy more credits or wait for your monthly billing cycle click over and give you more credits. That's how Clearstream works.

There's another one called Community, which is a little more expensive (that one's $100 a month). It works pretty much the same way, but it's more directly integrated with iOS, for iPhone users. There's a least a dozen of these services out there. Just make sure they focus on a few key things for whichever solution you go with.

- Really easy to use. Ideally, you or a volunteer should be able use it pretty effectively.
- Make sure it's not super stupid expensive. It should be something that is under a hundred bucks a month for you to message all of your people effectively.
- It should be easy to synchronize your current system with, not feel like a headache to get info imported to the service from your existing church database.

We've got to get on board with texting. Businesses, restaurants, and politicians are using it. Why can't we, the church, use it to connect with our people?

3 Minute Action Items:

1. Set up a free trial of Clearstream and start playing with it.

2. Research 3 different texting options if you don't like Clearstream

3. Schedule time in your next staff meeting to brainstorm a plan to leverage texting to communicate to your church.

THE GOSPEL & GENERATIVE AI (CHATGPT)

It's all the buzz right now, but this is not my attempt to latch onto some hype technology terms in order to sell books. The fact is, generative AI tools like ChatGPT are here to stay - and how we leverage them, approach them, talk about them, or disconnect from them will make a lot of difference in how we communicate.

Let's fly over some of the basic concepts of what this technology is, and then I want to dive into some theological framework and certain appropriate use cases for generative AI as ministers of the Gospel.

How big a deal is ChatGPT?

It's a big deal. Think internet-level disruption.

"ChatGPT is scary good. We are not far from dangerously strong AI," said Elon Musk, who was one of the founders of OpenAI before leaving. Sam Altman, OpenAI's chief, said on Twitter that ChatGPT had more than 1 million users in the first five days after it launched.

According to an analysis by Swiss bank UBS, ChatGPT is the fastest-growing app of all time. The analysis estimates that ChatGPT had 100 million active users in January, only two months after its launch. For comparison, it took nine months for TikTok to reach 100 million.

What is ChatGPT?

ChatGPT is a natural language processing tool driven by AI technology that allows you to have human-like conversations and much more with the chatbot. The language model can answer questions and assist you with tasks like composing emails, essays, and code.

Usage is currently open to the public free of charge because ChatGPT is in its research and feedback-collection phase. As of Feb. 1, 2023, there is also a paid subscription version called ChatGPT Plus.

Who made ChatGPT?

ChatGPT was created by OpenAI, an AI and research company. The company launched ChatGPT on Nov. 30, 2022.

OpenAI is also responsible for creating DALL-E-2, a popular AI art generator, and Whisper, an automatic speech recognition system.

Although ChatGPT is not the first or only AI tool of its kind, it's the most recognizable and most utilized.

Now, let's dive into some use-cases for pastors and ministry leaders when it comes to generative AI and how we use it to amplify the message of the Gospel and the Glory of God.

In 2013, I worked for a business in a shared open office space that a small church staff officed in as well. December of that year, Amazon released the first Echo device, their smart speaker with the Alexa technology. As a long-time Prime customer, I got on the early release list and was sent one of the first release models. I am a massive technology nerd, and it was exciting to see this breakthrough new technology, so I brought it to work one day to show off a bit.

We were using it to ask dumb questions like how tall is Mount Everest, complicated math equations, and things like that, when one of the older pastors of the church walked in from a meeting into the office. He stopped and gasped. “Is that that new Amazon device?!”

“Yeah, want to check it out?” I asked.

“Please unplug it and get it out of here. That thing is straight from Satan.”

I thought he was joking - I chuckled.

“I’m serious - we are ushering in the end times and the coming of the anti-Christ with technology like that”

I was a little taken aback.

I had my concerns about data collection, privacy, and Big Brother, but I did not see a connection between the anti christ and a "smart speaker".

Out of respect, I unplugged it and took it home. Months later, he left the church staff to move to another state and I brought Alexa back to the office.

At the risk of sounding similar to that, like a grumpy old man yelling at kids to get off my lawn, I want to ultimately warn of some concerns and grievances against using ChatGPT in certain ways.

I do not believe ChatGPT is a tool or gateway to the coming of the anti-Christ.

BUT

I have heard of pastors using it to write sermons, create ministry content, and even write Christian Non Fiction books under aliases, using ChatGPT as their personal ghostwriter. People I actually know personally are doing this, not just internet urban legends.

ChatGPT, Alexa, or any other technology breakthrough is not inherently evil or good. It's a tool. Like money or guns, the wickedness or goodness of the tool lies in who is using it and how it's being used.

Money can be used to build shelters for the homeless and fund global missions projects.

It can also be used to traffic drugs, build strip clubs, and fund pornography studios.

The money is not the problem. The user and use case are.

ChatGPT is a tool that can be helpful if it's used correctly, but I have a strong conviction that a lot of uses in ministry for it are straight-up plagiarism.

Writing a sermon using Chat GPT is taking a process that is meant to come from Scripture and the whisper of God, and outsourcing that holy endeavor to an open website crawling tool. You're stealing content from around the web, using ChatGPT as the hunting dog.

Ed Young, pastor of Fellowship Church in DFW and a massive voice for creativity in the church, has often famously said "God gave you eyes, so plagiarize". His intention in saying that is that there's nothing new under the sun, so as pastors and communicators, steal what you need to make your messages stronger.

No matter his intent, I massively disagree with that message. I just believe that we are short-changing the creativity of our God when we outsource ideas to other people or lift content from other places. On some level, this goes deeper than Chat GPT - for example, should we not borrow illustrations from Mike Todd we see on Youtube? Should we not download series kits from places like Ministry Pass or Download Youth Ministry?

Using tools like that to curate additional stories or illustrations to help add padding to the message is one thing.

Creating books, devotionals, and messages using generative AI is a whole other problem.

I can stand by this conviction because I have lived it.

In seven+ years of full-time ministry, I have borrowed one piece of sermon content ever. Even then, it was just the title, and I crafted the sermon series messages through prayer and study of the Word.

One time I was speaking at a church out of Acts 20. It's a unique story from scripture that the Lord had highlighted to me in my prayer time, and I had spent around 8 hours that week studying and preparing a message.

After the message, the lead pastor of the church came up to me and encouraged me with what a great message that was. "I've never heard a message on Acts 20 like that. Where did you get that from?"

I wasn't trying to be a smart alec, but I just said "the Bible"

He laughed. "Well I know that, but was it a message you saw from like Carl Lentz, or Mike Todd, or Matt Chandler or something?"

"No", I said matter-of-factly. "I didn't take it from anyone. I prayed, God highlighted that, and I studied and prepared to create the message myself. 100% Luke-Gajary-plus-Holy-Spirit-Original"

I was a little bothered by that interaction - not because I wanted the credit for what I felt like was a well-communicated truth, but because it was the first assumption from a seasoned church leader that something that good, that creative, MUST have been stolen.

It solidified in my heart an intense disagreement with what Ed Young said about plagiarism - that if the God we follow is as creative, wise, and infinitely loving as we say He is, why would we not emulate HIM only and trust His voice to be the leading source of our creativity?

That said, the reason you picked up this book is not for a theological dissertation on plagiarism and best practices for preachers. We've been talking about the church implementing technology to help expand the Kingdom in the online space, so let's land this section by exploring some good use cases for Chat GPT and other generative AI tools.

The summary of effective use as ministers of the Gospel should be to AMPLIFY our messages, not CREATE them.

Let's study a couple of real-world examples of how myself and the DigitalChurch team use ChatGPT in creative work for churches.

One of our designers and video editors messaged me in our Slack channel about how he had discovered how helpful it was to take transcripts of messages that a church pastor had preached and then pull key takeaways or "core truths" from the message. The designer could then take those quotes and create sermon quote graphics and short form videos for social media. I think his prompt was "Summarize the top 5 key points from this transcript" or something to that effect.

Recently, our team was tasked with creating a conference video opener about strong women of the Bible about a few key characters. We asked ChatGPT to take a few of our prompts and develop a script for a voiceover. It took a paragraph of prompts and generated a really helpful video script that got us about 70% of where we needed to be.

Another thing we have done is taken a sermon title or key points, and asked ChatGPT to generate 5 headlines based on the content that could be used for social media captions and Reels titles.

Do you see the pattern here? We use AI tools to automate, fill in gaps, and create some consolidated content from our original messaging. We don't use it to create the messaging- just amplify it.

By all means, use ChatGPT to enhance your ministry: however, establish your personal convictions about the heart of God and your role as His messenger, and allow that to guide to what extent the tools are used.

Some more Example prompts:

"Please give me 5 examples of why a Christian should pray"

"Using this text, create 5 headlines for social media posts: [Insert sermon text]"

"Create a catchy title for this sermon based on the idea of serving others sacrificially"

"Create 5 fun captions about coming to church this Sunday"

I hope you learn to leverage technological advances to better drive forward your church expansion and the building of the kingdom of God in your community. In that effort, set some boundaries and create a line in the sand to what you and your church will implement generative AI.

3 Minute Action Items:

Create a policy or position paper on how your church will use generative AI tools for ministry purposes. If you don't set a written standard, as young volunteers or staff join your teams, they will (rightfully so) push the boundaries of what they use AI to do for the church. Having clarity in a one-pager will help your church stick authentically to the convictions that your church has about the creativity of God and His Voice being the driver for ministry.

I didn't write this simple field guide to sell ten thousand books and make a boatload of money.

I wanted to develop a resource that gives you something actionable that you can start doing today to level up your church online.

Quick recap of what you can tackle next:

- Take a quick look at your website, put the vital information above the fold on the page, both on mobile and on desktop.
- Do an audit of your social media. Make sure your bio is about the user, not the church. Make sure it's about what's in it for them. Get good photography on there. Use Linktree or a similar service to help make things simple to load, so they don't have to jump through a bunch of hoops to get to your information.

Research, pray, and determine your position on generative AI in your approach to ministry. Clarify and communicate that position to your teams.

- Finally, get something in the texting arena to communicate with your people, to help keep people in the loop.

If you start implementing these keys, over the next few months you will massively level up your church online game without a ton of work, a ton of money and a ton of time.

I'd love to help you further if you have more questions. I created DigitalChurch.co to help the church win online, and if you don't have a creative team in your back pocket I'd love to help fill in the gaps for you.

You're spinning so many plates, you've got so much ministry you're doing, and you're just trying your best in the online space. And I know you're crushing it. But at the end of the day, God didn't call you to be a digital marketer. He called you to be a pastor.

I would love to talk to you more about what it would take to help Digital Church take some of the weight off your shoulders. When it comes to content creation, social media management, communications with your church, and all the stuff that I've talked about in this book, I can help you further.

Otherwise, I hope this was valuable to you in and of itself. If I can be of any more service to you or your church, please connect. We offer creative services to churches, and we also have resources, downloads, and training videos to further implement these theories on our website below.

Luke Gajary
http://Digitalchurch.co
Luke@digitalchurch.co